Ours is a big country.
Come along and explore.
We'll go from Portland to
Portland.

People in this country vote. They vote for members of Congress.

Congress meets here. Congress makes rules for our country.

Look at this part of the country. The sunshine is constant here! You can swim in summer. You can swim in winter too!

Trains cross the middle of our country. Here the farmlands seem endless. People grow lots of food on these farmlands.

The Arctic has lots of ice. Part of our country is in the Arctic. Don't be surprised to see icebergs.

Go west to see Mount Hood. It is near Portland. Mount Hood will impress you.

You can explore our country in this book. How would you describe this country of ours?